Dr. Abdul Rashid Dar
Prof. G. H. Dar

Conservation of Kashmir Himalayan Endemic Plants

Anchor Academic
Publishing

Dar, Abdul Rashid, Dar, G. H.: Conservation of Kashmir Himalayan Endemic Plants, Hamburg, Anchor Academic Publishing 2017

Buch-ISBN: 978-3-96067-110-7
PDF-eBook-ISBN: 978-3-96067-610-2
Druck/Herstellung: Anchor Academic Publishing, Hamburg, 2017

Bibliografische Information der Deutschen Nationalbibliothek:
Die Deutsche Nationalbibliothek verzeichnet diese Publikation in der Deutschen Nationalbibliografie; detaillierte bibliografische Daten sind im Internet über http://dnb.d-nb.de abrufbar.

Bibliographical Information of the German National Library:
The German National Library lists this publication in the German National Bibliography. Detailed bibliographic data can be found at: http://dnb.d-nb.de

All rights reserved. This publication may not be reproduced, stored in a retrieval system or transmitted, in any form or by any means, electronic, mechanical, photocopying, recording or otherwise, without the prior permission of the publishers.

Das Werk einschließlich aller seiner Teile ist urheberrechtlich geschützt. Jede Verwertung außerhalb der Grenzen des Urheberrechtsgesetzes ist ohne Zustimmung des Verlages unzulässig und strafbar. Dies gilt insbesondere für Vervielfältigungen, Übersetzungen, Mikroverfilmungen und die Einspeicherung und Bearbeitung in elektronischen Systemen.

Die Wiedergabe von Gebrauchsnamen, Handelsnamen, Warenbezeichnungen usw. in diesem Werk berechtigt auch ohne besondere Kennzeichnung nicht zu der Annahme, dass solche Namen im Sinne der Warenzeichen- und Markenschutz-Gesetzgebung als frei zu betrachten wären und daher von jedermann benutzt werden dürften.

Die Informationen in diesem Werk wurden mit Sorgfalt erarbeitet. Dennoch können Fehler nicht vollständig ausgeschlossen werden und die Diplomica Verlag GmbH, die Autoren oder Übersetzer übernehmen keine juristische Verantwortung oder irgendeine Haftung für evtl. verbliebene fehlerhafte Angaben und deren Folgen.

Alle Rechte vorbehalten

© Anchor Academic Publishing, Imprint der Diplomica Verlag GmbH
Hermannstal 119k, 22119 Hamburg
http://www.diplomica-verlag.de, Hamburg 2017
Printed in Germany

ACKNOWLEDGEMENTS

We are exceedingly grateful to the BGCI (Botanic Gardens Conservation International, London), HSBC (Hongkong and Shanghai Banking Corporation Limited) and NBRI (National Botanical Research Institute, Lucknow) for financial assistance. We articulate gratitude for the project staff, for their indispensable support for the successful finale of this project. Thanks are due to the Head, Department of Botany, University of Kashmir, Srinagar for his encouragement during the course of this study. The support and facilities rendered by some Centers and Departments, such as the Centre of Plant Taxonomy *(COPT)*, the Department of Botany, the Centre of Research for Development *(CORD)* of the University of Kashmir, and other Agencies, is highly acknowledged.

A. R. Dar & G. H. Dar

CONTENTS

CHAPTERS	PAGE NO.
1. INTRODUCTION	4-5
2. STUDY AREA	5-7
3. KASHMIR UNIVERSITY BOTANICAL GARDEN	8
4. OBJECTIVES	8
5. MATERIALS AND METHODS	8-10
6. RESULTS	10-24
7. DISCUSSION	24-33
6. LITERATURE CITED	33-35
8. PHOTOPLATES	36-42

1. INTRODUCTION

Endemic species (i.e., species with restricted ranges) have often been found to be concentrated in quite small areas irrespective of overall richness of taxa (Crisp *et al*., 2001; Kessler, 2002). Narrowly endemic species are by definition rare, and therefore potentially threatened (Myers *et al*., 2000). Thus, such species are given high priority in conservation strategies (Davis *et al*., 1997) because their small ranges render them particularly vulnerable to habitat loss (Balmford and Long, 1994) and because they are assumed more susceptible to anthropogenic habitat disturbance than widespread taxa (Kessler, 2001). Notwithstanding the debate regarding the effectiveness of using surrogate species for achieving conservation goals (Leroux and Schmiegelow, 2007), use of endemic species for understanding patterns of global biodiversity distribution (Lamoreux *et al*., 2006) and in identifying biodiversity hotspots (Myers *et al*., 2000) is considered as an effective approach to prioritize conservation efforts which has become inevitable due to a significantly high percentage (13%) of global flora threatened with extinction (Pitman and Jorgensen, 2002).

Among the many habitats, mountains were well known to harbour concentrations of endemic plants (Major, 1988) and elevational patterns of endemism appear to be influenced both by taxon-specific ecological traits (e.g. life form, reproduction, dispersal, demography, spatial population structure, competitive ability) in relation to their specific interaction with historical processes and by environmental factors, such as topographical fragmentation. The topographical dimensions of Kashmir with Siwalik at Jammu, Pir Panjal range, Valley of Kashmir and the Main Himalayan mass with Nanga Parbat, Upper Chenab valley, Zanskar range and cold desert of Ladakh merging into Tibetan plateau, serve as a crucible for the evolution of a flora of Pamirs and Hindukush, Karakoram mountains of central Asia, Ladakh-Tihetan flora and the main western Himalaya. Although a compendium of the extent of endemism in the Himalayas is not known but it is believed that majority of the 41.5% of dicot species endemic to India, are concentrated in the Himalaya (Dhar, 2002). Likewise in the Kashmir Himalaya, Dhar and Kachroo (1983a) reported 15.94% of endemics in monocots (Poaceae excluded) and an overall percentage of 31.38 endemic dicots. However, Dar and Aman (2003), reported occurrence of only 152 (*ca.* 8%) endemic taxa in this region, which constitute about 3% of the total Indian angiosperm endemics. However, it needs to be emphasized that the Kashmir Himalayan region constitutes only 0.48% of the total landmass of India, is geologically younger and, among adjoining regions, has the least area per endemic taxon. A significant proportion of these Kashmir Himalayan endemics (40%) are endangered due to a multitude of factors and several of them are now listed in the Red Data Books at the regional, national and international levels (Dhar and Kachroo, 1983b).

In spite of the reported high incidence of endemic taxa in the Indian Himalayan region, very little attention has been focused on the population profiling and identification of the threats impinging on them (Saharia, 1982; Dhar and Kachroo, 1983a and b; Hajara, 1983; Kapur, 1983; Dar and Naqshi, 1984; Maunder, 1988; Vir Jee *et al.*, 1990; Kaul, 1997; Dar and Naqshi, 2001; Dar *et al.*, 2006 a, b). This paucity of critical information necessitated the present study of explicating the macro-level factors constraining the populations of the target species through extensive field survey trips, and also to undertake measures for *ex situ* propagation and conservation of threatened endemic plant species so as to supplement their recovery and restoration under field conditions.

2. STUDY AREA

The Valley of Kashmir, often referred to as the 'paradise on earth' for its diverse rationale, is situated in the northern fringe of Indian subcontinent between 33°.20′ and 34°.50′ N latitudes, and 73°.55′ and 75°.35′ E longitudes, covering an area of about 16,000 sq. km. (**Fig.1**). The Valley is formed by a girdling chain of Himalayan mountains, namely the Pir Panjal Range of the Lesser Himalaya in the south and Zanskar Range of the Greater Himalaya in southeast to northeast and the west. It is believed that the Kashmir Valley was once a large lake called *Satisar*.

The entire territories of the Valley form two distinct topographic divisions, the mountain ranges and the Valley proper. The mountains vary in their height, rising up to an altitude of about 4,200m and are beset with sub-alpine and alpine meadows, and at the top with permanent glaciers. The Valley proper is an oval, alluvium-filled river basin with an altitude of about 1,600m at its capital city - Srinagar, and a variety of rich fresh-water bodies, such as springs, lakes, etc.

The climate of the Valley is predominantly temperate, changing to subalpine and alpine higher up in the mountains. A unique feature of the climate is four distinct seasons a year, namely spring (March–May), summer (June–August), autumn (September–November), and winter (December–February). The annual average precipitation is about 75cm, with sufficient rains during the months of March and April and also during July and August; during winter precipitation is mostly in the form of snow. July is the warmest month of the year, with temperature rising to an average of 29.5°C; January is the coldest month, with temperature coming down to -5°C. The maximum relative humidity (80 %) occurs during the months of November-December and the lowest (71 %) during May.

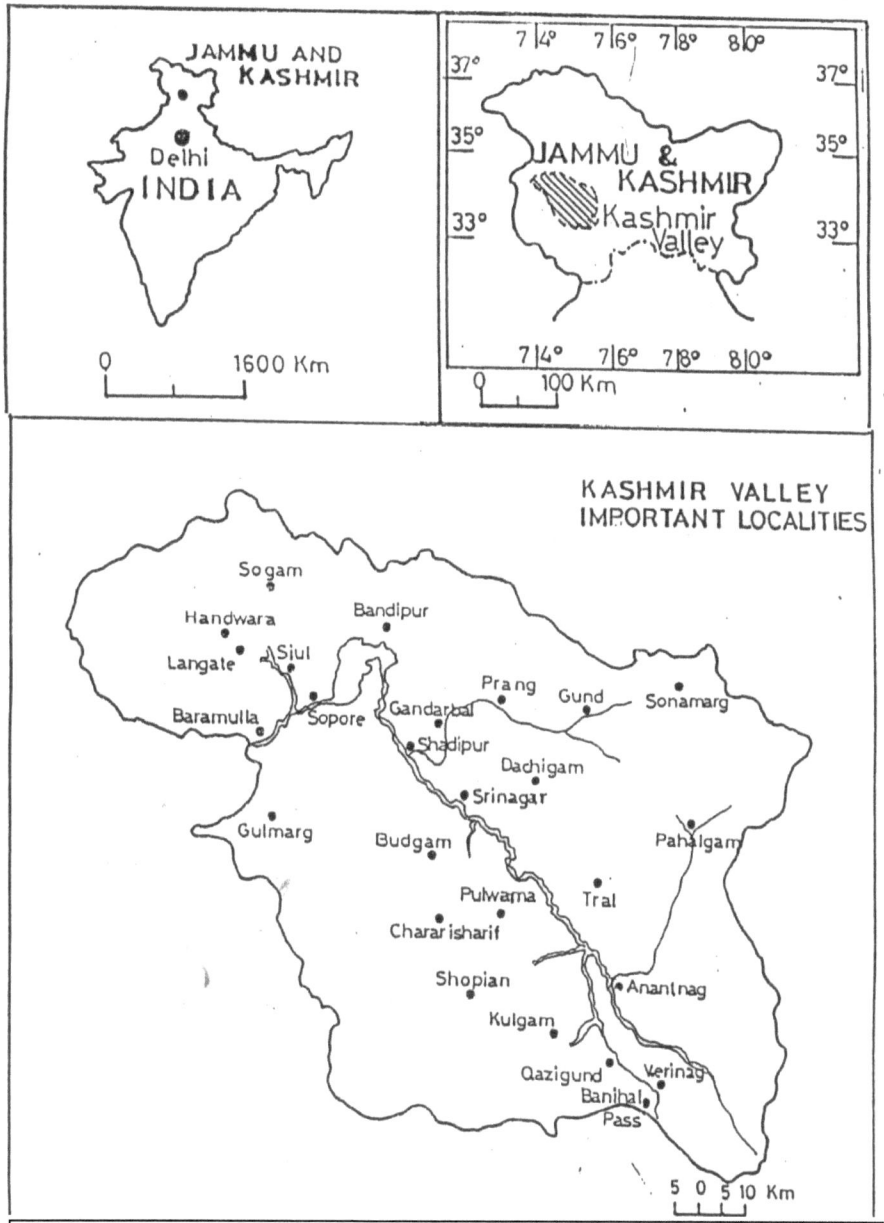

Fig. 1: Location of Kashmir Valley and its important localities

3. KASHMIR UNIVERSITY BOTANICAL GARDEN (KUBG)

Established in 1961, KUBG is situated at an altitude of 1,580 m (a. s. l.) and has an extension, the High Altitude Experimental Garden, at Gulmarg (alt. 2,500 m). Having a Glass house, Hot house, Pot house, and a Poly house, the garden is demarcated into different sections, such as Coniferatum, Shrubbery, Rosary, Salicatum, Deciduous-plant section, Bulbous-plant section, Rockery, Medicinal-plant unit, Lily pond and Canal, etc. These sections contain various kinds of plants for purposes of education, research, display, public awareness, and conservation. More than 500 plant species, mostly representatives of the local flora, presently grow in this garden. A large number of RET plants are also grown to afford them *ex situ* protection.

4. OBJECTIVES

i) To collect seeds/propagules of the target project species from their reported places of occurrence in the Kashmir Himalaya.

ii) To sow them in designated project plot at the Kashmir University Botanical Garden (KUBG) and standardize their propagation protocol.

iii) To enhance their propagation through various vegetative parts and seeds.

iv) To maintain populations of the target project species *ex situ* so as to supplement their recovery and restoration, and provide long-term back-up collections for their sustained use by the local populace.

v) To undertake measures for reintroduction of these species in to their natural habitats (e.g. High Altitude Botanical Garden, Gulmarg).

vi) To organize education and awareness programmes regarding biodiversity conservation for the students and general public.

5. MATERIALS AND METHODS
Preparation of land

A portion of land in the Kashmir University Botanical Garden (KUBG) was selected for the purposes of this project. In this experimental plot, many beds (approx. 8ft x 4ft) were prepared for growing the target endemic plant species.

Field surveys

A proper procedure was followed for organizing field survey trips. Detailed scanning of all the relevant literature, guidance of experienced persons, and knowledge from the local tribals were used to obtain relevant information regarding the distribution, appropriate sites, local names, etc. of the target species. Legal permission was requested from the concerned Govt. and military authorities with respect to the prevailing security scenario in the State. Equipped with the tents, collection tools, etc. field survey team was deported from the Kashmir University campus to the designated survey areas.

Specimen processing and taxonomic description

The target species collected, were identified and taxonomically described using the available literature and the specimens of these species (if any) available in the KASH. Usually a few specimens from each locality of occurrence of these species were collected and processed for herbarium purposes. The specimens were prepared into vouchers using standard herbarium methodology (such as pressing, drying, preservation, etc.). They were assigned specific numbers, depending upon the locality from where they have been collected. Species description, include author citation, diagnostic characters, habitat, and specimens examined.

Seed biology

Seed biology/germination studies on some species were undertaken. The seeds of *Lagotis cashmeriana*, *Meconopsis latifolia*, *Megacarpaea polyandra*, and *Saussurea costus* were put to several treatments, including scarification, pricking, distilled water, light, dark, and to various concentrations of gibberellic acid (2mM, 1.5mM, 1mM, 0.5mM, 0.25mM, 0.125mM); potassium nitrate (2mM, 1.5mM, 1mM, 0.5mM, 0.25mM, 0.125mM); kinetin (2mM, 1.5mM, 1mM, 0.5mM, 0.25mM, 0.125mM); thiourea (2mM, 1.5mM, 1mM, 0.5mM, 0.25mM, 0.125mM); as well as to different temperatures (10°C, 15°C, 25°C, 35°C).

Tissue culture experiments

Tissue culture studies were conducted on *Lagotis cashmeriana, Megacarpaea polyandra* and *Meconopsis latifolia*. The explant material (cotyledonary leaves, roots, etc.) of this trio of species was sterilized using Sodium hypochlorite solution (5 %) for 10-15 minutes, while their seeds were sterilized with 0.5 % Mercuric chloride for 3-5 minutes. The explant material was inoculated on Murashige and Skoog (MS) (1962) basal medium (full-strength). The bacterial contaminated material was given just a dip in 90 % alcohol and then reinoculated in the new MS basal full-strength medium. Some inoculated material of these species had to be discarded due to fungal contamination.

The sprouting of the planted material enabled us to exploit the young buds, leaves and petioles as explants in case of *Megacarpaea polyandra* and *Aquilegia nivalis* for tissue culture purposes. These explants after thorough pretreatment were inoculated in MS basal medium supplemented with Benzylamim purine (BAP), Auxins (IAA) separately, and in combination of different concentrations, i.e. basal medium containing BAP, basal medium containing IAA, and basal medium containing combination of BAP and IAA.

6. RESULTS

Many a possible habitats in the Kashmir Himalaya were extensively surveyed for the location and collection of the target endemic plant species. For the collection of these species the project team planned survey trips to those sites where from they have been reported previously. In all, 23 field survey trips were conducted from 2004 to 2006. During the course of these field trips, it became substantially evident that the project species are indeed extremely rare and very difficult to collect, due to their narrow distribution range and restriction to the inaccessible pockets in hilly terrains.

A total of 497 accessions of the target plant species have been collected, which on authentic identification were found to belong to nine species, namely *Aquilegia nivalis, Lagotis cashmeriana, Saussurea costus, Meconopsis latifolia, Megacarpaea polyandra, Hedysarum cachemirianum, Gentiana cachemirica, Aconitum kashmiricum,* and *Artemisia amygdalina*. These species mostly occupied fragile habitats with harsh edapho-climatic conditions and were represented by small, fragmented populations (**Table 2**). These species were maintained as the *ex situ* collections in the selected plot at the KUBG.

Voucher specimens of all the collected species have been prepared and deposited in the Kashmir University Herbarium (KASH). Diagnostic and lucid morphological descriptions were prepared for all of these species.

The propagules/seeds of all the collected species were sown in the beds prepared for this purpose at the KUBG. Success has been achieved in the *ex situ* survival of all these species, except *Meconopsis latifolia*, in our Botanical Garden where they sprouted successfully in spring. Their propagation under the *ex situ* conditions has been better effected through the underground perennating organs - rhizomes/roots. In case of *Meconopsis latifolia*, however, the root is too soft and fragile, and only one individual sprouted. The seeds of all these species seem to be hard to germinate, requiring some specific treatments under the *ex situ* conditions. Propagation of some of these species through seeds, and a few of them through tissue culture techniques, was also attempted.

Target species

The nine critically endangered endemic species selected for the purposes of this project are:

TARGET SPECIES	
i) *Aquilegia nivalis* Falc. ex Baker	vi) *Hedysarum cachemirianum* Benth. ex Baker
ii) *Lagotis cashmeriana* (Royle) Rupr.	vii) *Gentiana cachemirica* Decne.
iii) *Saussurea costus* (Falc.) Lipsc	viii) *Aconitum kashmiricum* Stapf. ex Coventry
iv) *Meconopsis latifolia* Prain	ix) *Artemisia amygdalina* Decne.
v) *Megacarpaea polyandra* Benth.	

Seed collection

Great efforts were made to collect seeds of the project species for the purposes of sowing them in the designated area at KUBG. This was really a painstaking exercise, during which seeds of four species, namely *Saussurea costus*, *Meconopsis latifolia*, *Lagotis cashmeriana*, and *Megacarpaea polyandra* were collected at some places.

Sowing of the collected plant material

The project plant species collected were planted in different beds in the selected plot. Watering, deweeding, and protection from disease and rodents was being paid utmost attention. For watering of the project plants, a fifty-foot deep tube well was dug in the project plot itself. This greatly eased out the problem of watering the planted material in the said plot and aided in their survival.

Table 1: Details of field survey trips to collect the target species during the first year (May 2004 to May 2005)

Locality	Date	Species collected	No. of accessions collected	Part collected
Gulmarg (Apharwat to Sarsoon Post)	10-13th June 2004	*Aquilegia nivalis*	25	*Whole plant*
Sonamarg (Baltal)	25-29th June 2004	*None*	-	-
Gulmarg (Apharwat)	10 -14th July 2004	*Lagotis cashmeriana*	04	*Whole plant*
Uri (Ban Behak)	5-11th Aug. 2004	*Saussurea costus*	15	*Rhizome/ fruiting heads*
Auro (Rayil)	29th Aug.-4th Sep. 2004	*Meconopsis latifolia*	10	*Whole plant / capsules*
Sonamarg (Thajwas)	7-10th Sept. 2004	*Meconopsis latifolia*	20	*-do-*
Uri (Gagargalli)	15 -21st Sep. 2004	*Megacarpaea polyandra*	6	*Rhizome / whole plant with fruits*
Baltal to Amarnath	28th Sep – 6th Oct. 2004	*None*	-	-
Ban Behak (Uri)	26-28th March 2005	*Artemisia amygdalina*	20	*Rhizome*
Dara (Harwan)	29th March 2005	*None*	-	-
Naranag (Kangan)	31st March 2005	*None*	-	-
Gund (Sind Valley)	18th May 2005	*None*	-	-

Table 2: Details of field survey trips to collect the target species during the second year (June 2005 to June 2006).

Locality	Date	Species collected	No. of accessions collected	Part collected
Sonamarg (Hangduph)	26th May – 3rd June 2005	-	-	-
Naranag	12th – 19th June 2005	*Megacarpaea polyandra* *Meconopsis latifolia*	13 06	Rhizome Seedlings
Gulmarg (Khillanmarg)	29th June - 4th July 2005	*Aquilegia nivalis,* *Lagotis cashmeriana,* *Aconitum kashmiricum*	07 18 66	Whole plant Whole plant Whole plants / tubers
Naranag	13th -22nd July 2005	*Saussurea costus* *Lagotis cashmeriana*	17 13	Rhizome Whole plant
Sonamarg (Booj ki Lakd, Hapathgandh, Zaildardub)	29th July – 11th August 2005	*Lagotis cashmeriana* *Aquilegia nivalis*	55 09	Whole plant Whole plant
Naranag	17th – 27th August 2005	*Megacarpaea polyandra* *Hedysarum cachemirianum*	09 11	Whole plant Whole plant
Ladakh	29th August – 13th September 2005	-	-	-
Gangabal (Harmukh)	21st-29th September 2005	*Aquilegia nivalis* *Gentiana achemirica*	13 65	Whole plant Whole plant/ rhizomes/ fragmented rhizomes
Gangabal	17th – 16th October 2005	*Gentiana cachemirica* *Hedysarum cachemirianum* *Saussurea costus*	28 04 09	Whole plant/ rhizomes Whole plant Rhizomes.
Gulmarg (Apharwat)	22nd – 29th October 2005	*Aconitum kashmiricum* *Lagotis cashmeriana*	27 17	Whole plant/ tubers Whole plant
Gangabal	26th May to 3rd June 2006	*Meconopsis latifolia*	11	Seedlings

From the above **Tables 1 & 2**, it can be concluded that the three target species which could not be located during the first year, were collected in the first-half of the second year; besides many accessions were added to the already collected target species. The 4th project species collected during the summer season of the year 2005 was *Artemisia amygdalina*.

From May 2004 to June 2006, a total of 497 plant accessions of the target project species were collected in the form of various propagules. Out of these, 99 accessions of 6 species were collected in the first year, and 398 of 9 target species in the second year of project tenure. The 9 target species collected and planted so far have shown good results. Out of these, *Artemisia amygdalina, Sassurea costus, Aconitum kashmiricum, Gentiana cachemirica, Hedysarum cachemirianum*, and *Lagotis cashmeriana* showed good results in their survival and establishment.

Threat status and its field assessment

All the target species have been collected from different areas in the form of 497 accessions (**Table 1, 2**). The populations of all these species are small and quite fragmented in nature; the data pertaining to their threat status assessment is given in **Table 3**.

Table 3: Threat status assessment of the target project species

Name of species	Threat status	Current threats	Threat status assessment in the field
Aquilegia nivalis	CR	Fragile habitats, grazing pressure	Very small, restricted populations, growing in extreme alpine conditions
Lagotis cashmeriana	CR	Fragile habitats, grazing pressure	Small, fragmented populations, with low sexual reproductive potential
Saussurea costus	CR	Heavy exploitation for medicinal purposes, illicit trade and grazing pressure	Reduced, restricted populations, less sexual reproductive effort, only a few viable seeds per head, overexploitation of rhizomes greatly affecting its vegetative propagation
Meconopsis latifolia	CR	Extreme fragile habitats, very rare local endemic	Highly shrunk populations, extremely fragile nature of plant especially root-stock affecting vegetative propagation
Megacarpaea polyandra	CR	Harsh alpine habitats, exploitation as a vegetable by the local tribals	Extremely restricted and shrunk populations, only a few reproductive individuals observed
Artemisia amygdalina	CR	Sandy foothills in moderately moist, subalpine habitats	Populations restricted to an isolated area in the Lower Jhelum Valley; flowering tops browsed by the cattle
Hedysarum cachemerianm	CR	Extremely vulnerable, typical open alpine slopes	Squeezed populations, very restricted distribution in few alpine areas, least number of flowering individuals and meager seed production
Gentiana cachemirica	CR	Crevices of huge sedimentary rocks with extremely scarce soil	Extremely harsh and fragile alpine habitats with highly specific ecological niche
Aconitum kashmiricum	CR	Loose-soiled, less-pebbled, moist, extremely fragile, open alpine slopes	Harsh and fragile habitats, squeezed populations and restricted distribution, extensive grazing of flowering portions, exploitation of tubers for medicinal purposes

Taxonomic description

The species collected have been morphologically described; the description includes author citation, diagnostic characters, habitat, common name and specimens examined, as follows:

a) *Aquilegia nivalis* Falc. ex Jackson (*A. vulgaris* subsp. *jucunda*) Hook. f. & Thomson in Hook. f., Fl. Brit. India. 1: 24. 1872.

Perennial herb, attaining 10-25 cm height; stems solitary or rarely few from the base, short, usually leafless; root deep, solitary, cylindrical, thick, slightly spongy, blackish-brown. Radical leaves few (2-10), petiolate, bipinnate; petiole 2-8 cm long, with clasping base, slightly flattened, finely and longitudinally grooved, pubescent; leaflets subsessile, kidney-shaped, 3-lobed; lobes broad, blunt, pubescent. Cauline leaves usually absent. Flowers solitary, terminal, drooping, 2-5 cm across, deep purple; sepals 5, petaloid, ovate, slightly spreading, blunt; petal limb about equaling or slightly longer than stamens; spur conic–obtuse, long, straight or slightly curved.

Habitat: Occurs in moist shady or open places, in slightly hard or pebbled soil at an altitude of 3,000 – 3,800 m. Common name: Columbine. Specimens examined: Gulmarg, 3800 m, 10.06.2004, *G. H. Dar* and *A. R. Dar* 01 (KASH); Brarimarg (Baltal-Amarnath route), 3400 m, 21.08.2004, *G. H. Dar* and *A. R. Dar* 11 (KASH).

b) *Lagotis cashmeriana* (Royle) Rupr. Mem. Acad. Sci. st. Petersb. VII, 14 (4): 64. 1869.

Fleshy, perennial herb; stem usually solitary, ascending, 7-25 cm long, amidst radical leaves, scape-like, leafy above; roots shallow, profusely-branched, cylindrical, spreading, thin, shallow, fleshy, moderately hard, reddish–brown. Radical leaves obovate-elliptic, petiolate (petiole 1-4 cm long), simple, leathery to touch, 2-5.5 x 1-3 cm, appearing enclosed within a common sheath-like structure at their origin, entire, crenate, acute; cauline leaves semi–amplexicaul, sessile, ovate, 0.5-3 x 0.5-1.5 cm, acute, serrulate. Flowers arranged on terminal (3.5-12 cm long) portion of flowering spike; each flower 0.5-1.3 cm long, dark-blue; calyx dorsally plane, lobes fimbriate; corolla prominently curved, upper lip entire or 2-3 toothed, lower 2-lobed; anthers reniform. Seeds ovate, pointed, hard, light black-brown.

Habitat: Occurs in wet places (glacier-fed) in open or shady areas, rock crevices, loose-soiled and less-pebbled patches at an altitude of 3,000-4,000 m. Common name: Kashmiri- Hare's Ear. Specimens examined: Gulmarg, 3800 m, 11.06.2004, *G. H. Dar* and *A. R. Dar* 02 (KASH); Thajwas, 3150 m, 19.08.2004, *G. H. Dar* and *A. R .Dar* 08 (KASH); Rayil, 3200 m, 25.08.2004, *G. H. Dar* and *A. R. Dar* 12 (KASH); Apharwat (Gulmarg), 3500 m, 24.07.05, *G. H. Dar* and *A. R. Dar* 10 (KASH); Naranag,, 3600 m, 18.07.05, *G. H. Dar* and *A. R. Dar* 13 (KASH).

c) *Saussurea costus* (Falc). Lipschitz in Bot. J. URSS 49: 131. 1964 et Rod Saussurea 92. 1979.

Tall robust perennial herb, 0.5-2.5 m tall; stem simple, pubescent; rhizome deep, cylindrical, thick, less-branched, blackish-brown. Basal leaves petiolate, triangular, 25-50 x 15-20 cm, irregularly toothed, pubescent; petiole lobately winged; cauline leaves short-petiolate to sessile, semi-amplexicaul, 10-20 x 8-15 cm. Flower heads terminal or axillary, solitary or clustered (2-7); each head sessile, subglobose, 2-4.5 cm across, hard, purple; florets dark-blue to purple; involucre numerous, ovate–lanceolate, rigid, recurrved, hairless, purple; receptacle with long bristles; corolla dark-purple. Cypsela curved, compressed, 4-9 x 2-3 mm; pappus brown.

Habitat: Occurs on moist, shady slopes among juniper shrubs or in open places at an altitude of 2,800-3,800 m. Common name: Kuth. Specimens examined: Gulmarg, 3100 m, 15.06.2004, *G. H. Dar* and *A. R. Dar* 05 (KASH); Banbehak (Uri), 3200 m, 5.08.2004, *G. H. Dar* and *A. R. Dar* 06 (KASH); Naranag,, 3100 m, 18.07.05, *G. H. Dar* and *A. R. Dar* 14 (KASH)..

d) *Meconopsis latifolia* (Prain) Prain in Bull. Misc. Inf. Kew 1915: 146. 1915

Herb, monocarpic; stem simple, 15-50 cm tall, leafy, spongy, delicate, finely-grooved, covered with minute, sharp, golden-brown, 2-7 mm long bristles; tap root deep, less branched, fleshy, brittle, slightly whitish. Basal leaves petiolate, oblong or ovate–lanceolate, 4-23 x 1.5-7 cm, acute, undulate, covered on both surfaces by golden-brown, 2-6 mm long bristles; upper leaves subsessile to sessile. Flowers axillary or terminal, pedicellate, many in a spike-like cluster borne on leafy stem, uppermost flower usually ebracteate; sepals broadly-oblong, 1-2 cm long; petals 4, broadly-ovate, 2-3 x 1.5-2 cm, light-blue, obtuse, with slightly wavy upper margin; stamen filaments deep-blue, anthers orange-yellow; ovary ovoid, bristly. Capsule obconic or oblong, dehiscing by 4-7 valves, with stylar beak, and capsule covered with sharp, slightly harder bristles. Seeds reniform, finely grooved, brown, covered all over with minute golden-brown bristles.

Habitat: Occurs in rock crevices with thin soil, under rocks, or among big boulders in sandy soils at an altitude of 2,800 – 4,000 m. Common name: Blue poppy. Specimens examined: Najwan (Badmarg), 2900 m, 15.07.1983, *G. H. Dar* 6876 (KASH); Banbehak (Uri), 3200 m, 5.08.2004, *G. H. Dar* and *A. R. Dar* 07 (KASH); Thajwas, 3000 m, 25.06.2004, *G. H. Dar* and *A. R. Dar* 03 (KASH); Naranag,, 3750 m, 14.07.05, *G. H. Dar* and *A. R. Dar* 17 (KASH); Gangabal, 3850 m, 21.09.05, *G. H. Dar* and *A. R. Dar* 17 (KASH).

e) ***Megacarpaea polyandra*** Benth. In Hooker's J. Bot. Kew Garden Misc. 7: 356, t.7.1855.

Robust perennial herb or undershrub; stem up to 80 cm tall, much branched, glabrous; root much deep, thick, least branched, cylindrical, tapering towards tip, fleshy, whitish. Basal leaves petiolate, bipinnate, 20-50 cm long; leaflets lanceolate, 8-15 x 2-3 cm, serrate, acuminate; petiole auricled at base, flattened, spongy. Racemes panicled, many-flowered; flowers creamy-yellow, upto 1 cm across, pubescent; sepals obovate or elliptic–obovate; petals rounded, stalked, creamy-yellow; stamens 8-16. Fruits suborbicular, 3.5–5 cm broad, one-seeded, bilobed, lobes unequal in size, sometimes one lobe abortive. Seeds reniform, creamish-yellow, flattened.

Habitat: Occurs in open, moderately moist, loose soils. Common name: Chatter. Specimens examined: Gagargalli (Uri), 3600 m, 8.08.2004, *G. H. Dar* and *A. R. Dar* 09 (KASH): Naranag, 3800 m, 14.07.05, *G. H. Dar* and *A. R. Dar* 15 (KASH).

f) ***Hedysarum cachemirianum*** Benth. ex Baker in Hk. f., FBI 2: 146, 1876.

Robust, hairless, erect, 30-60 cm tall, perennial, undershrub, with dense cluster of numerous, large, drooping, red-purple flowers borne on nearly leafless stem. Leaves imparipinnate, 7-18 cm long, 19-27-foliolate; leaflets linear-oblong, 1.2-2.5

cm long, obtuse, ultimately glabrescent on either face. Flowers very shortly pedicelled, axillary, 2-2.5 cm long, in compact racemes (4-7.5 cm). Calyx 0.9-1.1 cm long, downy, with 5 narrow, acuminate teeth, pubescent, partly coloured. Petals 5, free unequal, much exerted from the calyx, the standard longer than the wings and keel. Stamens 10, diadelphous. Carpels united; ovary linear and stalked; style very long, filiform. Fruit, a pod with 1-3 flattened, oblong, indehiscent joints.

Habitat: loose, moderately- moist, open alpine slopes at (3,200-4,000 m). Common name: Kashmir Hedysarum. Specimens examined: Naranag, 3680 m, 17.07.05, *G. H. Dar* and *A. R. Dar* 16 (KASH).

g) *Gentiana cachemirica* Decne. in Jacq. Voy. Bot. Inde. 111, t. 17, 1844.

Tufted, glabrous, leafy perennial: stem 5-6 cm long prostrate or ascending from the root; branches often many from the root, 8-15 cm long. Leaves broadly elliptic or oblong, crowded, opposite, 6-15 x 0.4-0.8 cm, narrowed into short petioles, obtuse, apiculate, the radical ones spathulate or obovate-oblong, sharply pointed. Flowers bright blue, 2.5-3 x 2.5 cm, solitary, terminal, sessile. Calyx tube, long, bell-shaped, oblong, with a wide sinus between the lobes. Corolla lobes broadly ovate or elliptic, acute, with erect, triangular, unequally fimbriate lobules in the sinuses. Stamens 5, inserted on the corolla tube. Ovary, cylindrical, with a short, straight style. Fruit, a small narrow capsule.

Habitat: crevices of huge sedimentary rocks with extremely scarce soil, rarely in soil near such rocks at (3,200- 3,800 m). Common name: rock Gentian. Specimens examined: Naranag, 3500 m, 11.09.05, *G. H. Dar* and *A. R. Dar* 16 (KASH).

g) *Aconitum kashmiricum* Stapf ex Coventry, Wild Flrs. Kashmir 3: 25, t.   13.1930

Erect, tuberous herb; stem 10-20 cm high, erect, slender, simple or often branched at the base, and sparingly branched above, glabrous or pubescent on the upper part. Leaves, few, glabrous, all long-petiolated except one or two on the upper part of stem, which are sub-sessile; the blade orbicular-cordate or ovate-cordate, 1-3 cm in diameter, incised into broad lobes with 3-5 apiculate teeth. Flowers 1.5-3.5 cm in length, solitary, or two together at the top of the stem, or in racemes of a few distant flowers; pedicels erect, pubescent, 1-3.5 cm in length. Sepals 5, petaloid, upper one helment shaped, with a small crest near the tip. Petals 2, small, reduced to nectaries enclosed in helment and consisting of a slender, curved, channeled claw with a small limb. Stamens, numerous, filaments hairy on upper part and winged below. Carpels, 5, free, densely pubescent. Fruit, a cluster of 5 free, erect, hairy follicles.

Habitat: loose-soiled, less- pebbled, moist and open alpine slopes at (3,000-3,800 m). Common name: Pevak, Kashmir Mankshood. Specimens examined: Gulmarg 3500 m, 11.09.05, *G. H. Dar* and *A. R. Dar* 19 (KASH).

h) *Artemisia amygdalina* Decne. In Jacq. Voy. India. 92. t. 100. 1844.

Perennial herb, large, glabrous; rhizome solid, hard, woody, 2.5-3.5cm in diameter, blackish brown, almost straight–slightly curved; stem erect, 2-3 m in height, 0.5 – 1.5 cm in diameter, leafy, prominently grooved, bears dense, small, leafy branches and branchlets. Leaves simple, rather membranous, acuminate, glabrous and bright green above, dull white and pubescent below, serrate, 3-9cm in length and 0.5 – 1.5 cm in breadth, sub sessile-sessile. The terminal, tapering portion of stem bear number of flowering heads; flowering heads in terminal branched raceme, ovoid, 0.5-1.5 cm in length and 0.4 – 0.6cm in diameter; involucre bracts oblong, obtuse, glabrous with pappery margins; outer florets, female, fertile; disc florets hermaphrodite, fertile, yellow, tubular.

Habitat: moderately- moist, relatively hard, open or partially shaded sub- alpine situations at (2,600-3,200 m). Common name: Veri tethvan. Specimens examined: Uri, 2950 m, 28.03.05, *G. H. Dar* and *A. R. Dar* 18 (KASH).

Usefulness of the target species

As most of these target project species are narrow local endemics and as no precise studies have been made on their various aspects, the information about their medicinal, ethno medicinal and other values is scarce. Only scattered information is available about a few of them.

• *Megacarpaea polyandra*

Fleshy roots are relished as pot herb or eaten raw. Leaves are prized as spinach.

- *Meconopsis latifolia*

Roots powdered and eaten with water as stomachic.

- *Saussurea costus*

Rhizome used as spasmodic in asthma, cough and cholera, in skin disease and rheumatism, also as insect repellent.

- *Lagotis cashmeriana*

Whole plant considered medicinal. The paste of leaves applied for wound healing in cattle. Rhizome used as adulterant in place of *Picrorhiza kurrooa*.

Standardization of propagation through vegetative propagules

The afore-said plant species were collected in the form of whole plants, rhizomes, or seeds. The propagules sown in the project plot were pursued from sowing till their sprouting afresh in the spring. It was found that 8 out of the 9 species yielded best results in terms of the rhizomes, whereas the in case of the 9th species- *Meconopsis latifolia*, only one individual sprouted from its roots/rhizomes. In contrast, seeds did not show such encouraging results in terms of their germination both in the field beds and *in vitro*. Thus, rhizome has been ascertained to be the best vegetative propagule for the survival, healthy sprouting and sustenance of most of the project species maintained.

Development of protocol for optimum seed germination under *ex situ* conditions at KUBG

The collected seeds of *Meconopsis latifolia*, *Lagotis cashmeriana*, *Megacarpaea polyandra* and *Saussurea costus* were divided into two sets each. One set of seeds of these species was sown in separate beds in the project plot at KUBG; whereas the other set of each was subjected to various treatments in the laboratory as mentioned in the methodology. The seeds sown in the field beds did not germinate at all, except in *Megacarpaea polyandra*, where also out of the forty seeds sown only seven seedlings have turned out. The seeds of *Lagotis cashmeriana* and *Meconopsis latifolia* were subjected to laboratory trials. In these laboratory experiments, seeds of *Meconopsis latifolia* have shown good results with respect to various concentrations (1mM, 0.5mM, 0.25mM) of gibberellic acid and potassium nitrate, while only a few seeds of *Lagotis cashmeriana* have germinated in various concentrations (2mM, 1.5mM, 0.5mM) of gibberellic acid and thiourea. Particularly, seeds of *Meconopsis latifolia* were specially treated to obtain some seedlings, so as to establish this plant species in KUBG (which otherwise did not sprout from the vegetative material collected from the field). This also provided us the explant material for tissue culture purposes of this fragile species. In case of *Saussurea costus* the *in vitro* as well as the *ex situ* trials showed zero results, mostly because the seeds seem to be non-viable. The seeds of *Megacarpaea polyandra* collected were not mature enough at the time of collection. In fact we collected twenty-six specimens of the said species, of which only one bore seeds (non-harvestable). We collected this seeded specimen as such and planted it in the project plot with the hope that as the upper part of the plant dries up, some resources may be allocated towards these immature seeds so that we would get a few viable seeds. Our plan did work and we got a few viable seeds. These seeds when kept under observation in petridishes in distilled water at a temperature of 20-25°C, germinated slowly without any further special treatment. The seedlings thus obtained were used for tissue culture and some transferred to trays kept in the polyhouse during winter. In tissue culture experiment, few inoculated seedlings showed initial survival and growth, but ultimately were lost due to contamination; whereas few seedlings in trial pots survived and they sprouted afresh in the spring.

Tissue culture studies

A few leaf explants of *Lagotis cachemiriana* showed encouraging results; even one leaf explant developed a healthy callus. In case of *Meconopsis latifolia,* the explant (cotyledonary leaves and roots) dried after few days of inoculation, and a few seedlings we had obtained from seeds unfortunately got lost due to repeated contamination. Few seedlings of *Megacarpaea polyandra* also met the same fate. We also inoculated the seeds of *Meconopsis latifolia, Lagotis cachemiriana* and *Megacarpaea polyandra* on the same basal MS medium. Out of all these seeds, only one seed of *Megacarpaea polyandra* germinated and developed into a full seedling; however, the seeds of *Meconopsis latifolia* and *Lagotis cachemiriana* failed to germinate. The sprouted material of the *Megacarpaea polyandra* and *Aquilegia nivalis* in the project plot was also utilized for inoculation in the freshly prepared medium. Almost all of these explants, only after few days of inoculation had to be discarded due to fungal contamination.

7. DISCUSSION

The project team made well planned and target oriented, extensive field survey trips to the difficult and inaccessible subalpine and alpine terrains throughout the Kashmir Valley. These efforts eventually succeeded and all the target species were collected.

In terms of accessions, many plant specimens of the already collected species (2004-05) were supplemented during the second year (2005-06); in addition, large numbers of plant accessions were also collected for the uncollected species. As per availability in the field, mostly whole plants or rhizomes of these plant species were collected, packed in polythene bags at the field site, and immediately transported to the project plot at the KUBG. The collected propagules were planted in the beds at KUBG and kept under constant supervision.

A critical analysis of the planted material of the target species at KUBG vis-à-vis their natural habitat conditions served as a very helpful tool in devising a renewed experimental setup for their maximum survival, growth, vigour, reproduction and seed set attributes. For this purpose same species accessions were divided into many subsets and planted under different conditions, e.g. in the open, shade and altered soil conditions in the field beds as well as pots. The results showed variations in terms of survival, vigour and to some extent in reproduction.

The analysis of the results regarding the survival, growth, etc. of these target species planted at KUBG revealed that the overall outcome was very much up to the satisfaction of the project team. Species-wise, *Artemisia amygdalina* showed exceptionally best results. The species successfully multiplied by the underground rhizome and the individuals attained more height as compared to those in the natural populations.

In this plant species, the collected rhizomes planted in the open sunlight flourished well in the experimental plot, the latter is now densely covered by the rapid spread of these individuals. The 20 plant accessions which were collected originally from the field and then planted in the KUBG, multiplied into a large number of individuals. From a single piece of rhizome, which at the time of collection just bore only one off-shot in its natural habitat, about 3-6 off-shoots were produced in the experimental plot. These off-shoots grew vigorously in the plot and attained a height of 2-3 m; and even produced inflorescence in the very first year of their establishment. It seems to be an unusual phenomenon as in the perennial plants, at the initial stage of their life history most of the plant's resources are used to survive and acclimatize in the transplanted habitat. Usually, such plants take at least 1-2 years to first get established and then allocate resources for the reproduction; even then only some species produce flowers and seeds. To achieve the acclimatization accompanied with vigorous growth, propagation, reproduction. On comparing growth of this species in the experimental plot and in its natural habitat, it was found that individuals attained greater height (2-3 m) and vigorous growth in the former than in the latter, where height was between 1-2 m and growth was normal.

With an established and healthy plant material of *Artemisia amygdalina* present in the experimental plot, we attempted to enhance propagation of the species at a larger scale. Juvenile leafy shoots were excised carefully from the main stem and transferred to an already designed experimental setup consisting of two sets of trial pots: one set containing normal garden soil and second containing soil and sand mixed in the 1:1 ratio. The experiment aimed to promote rooting of the off-shoots. The juvenile offshoots were planted in these trial pots and watered regularly, some trial pots were placed in the open and some in the shade. Some of the offshoots were also planted in the open and shaded field beds in the project plot. The entire experimental setup was kept under constant observation. After about 20-25 days, several planted offshoots in each plot were dug out carefully and checked for rooting. It was found that in each case a few whitish slender and thin roots had developed on the underground portion. On comparison, the individuals in the shaded trial pots and beds showed less pace of growth as against those in trial pots and beds kept in the open sunlight. The leaves in shaded plants were pale-green colored and their stem relatively weak as compared to open planted offshoots. Similar experiments were performed to stimulate the stem cuttings to root and actually accelerate the already started mass propagation.

With the success of experimental trials, we now have enough material available to attempt for the reintroduction of this species (in the form of different types of propagules) at selected high altitude sites in the Kashmir Himalaya, starting first with reintroduction in the protected sites (e.g. High Altitude Botanical Garden-Gulmarg, altitude 2500 m); at the latter site, the species is growing successfully. It seems most probable that this species will not face any problem, if transferred to its natural habitat along high altitude sites in the Kashmir Himalaya.

In the natural habitat of the *Artemisia amygdalina*, the population was observed continuously for a few months in the spring and summer seasons (2004-2005). At the beginning, it appeared that most of the individuals bore no inflorescences, thereby probably impacting its population size. However, on further investigation, it became quite evident that the overall population density was higher and there appeared no signs of decline in the number of individuals. In addition, it was found that the habitat is visited regularly by tribal people, whose cattle graze only on the inflorescence of flowered individuals and no other part. Such de-topping of its vital reproductive organ has continued over decades and led to the plant's inability to rely on propagation through seeds. To compensate, the plants have started investing more resources to their rhizomes and developed them as potential and effective means of propagation. Still propagation by seeds has not been stopped altogether, it appears to have been kept as a subsidiary means, although most of the seeds produced through sexual reproduction have been found to be non-viable. During the long course of time, perhaps the plant species may stop allocating resources towards the seed production if it finds that the prevailing disturbing conditions do not improve.

Artemisia amygdalina, last located in 1971, was altogether neglected since then. This narrow endemic species may prove potentious for this region and to overall humanity as is evident from its well known ethno-botanical importance and medicinal properties of other related species of the genus. Over the years, many operational factors, such as low population size, habitat specificity, narrow distribution ranges, land-use disturbance, heavy livestock grazing, construction of dams and roads, fragmentation and degradation of habitats and population bottle-necks (Kala 1998, 2000; Weekly and Race 2001; Oostemeijer *et al*. 2003; Verger *et al*. 2003) have drastically reduced the populations of rare species. The *A. amygdalina* is no exception to this phenomenon. The location and collection of this species from the Kashmir Himalaya was looking for a proverbial needle in a haystack. Most often, exertive and troublesome field survey trips proved mere failures. This may be evident from the fact that in the present study the said species was collected from only a single locality in the Kashmir Himalaya till date. The successful cultivation of the species in the experimental trial resulted in devising standard methodology for its mass-level propagation from various organs through vegetative means, a unique feat achieved for this species. The methodology is easier one, less-sophisticated, adaptive, economical, time saving and above all practical as compared to the usual modern methods (e.g. tissue culture) used for the propagation of RET species.

In terms of survival and growth attributes, the *Saussurea costus* followed next to *Artemisia amygdalina*. This species showed maximum survival, as most of the rhizomes planted sprouted and showed vigorous vegetative growth and some individuals even produced flowers with few viable seeds. The rhizomes were carefully watered, as excess watering could have resulted in their rotting, and most of them sprouted in the consecutive years.

The *Aconitum kashmiricum*, a typical alpine species exclusively restricted to loose soils, was collected in the growing season of the year 2005. The species, at its natural sites, was represented by a good number of individuals. The 93 plant accessions collected during the project period were planted in different experimental setups at the KUBG, such as in the project plot beds, pots etc. About 75 accessions of this species survived, acclimatized, flourished and a few of them produced flowers and seeds. The species found the pots with moderate levels of moisture, little shade and the soil mixed with sand as the best condition for survival, growth, flowering and multiplication. During 2006 almost all the individuals survived and sprouted in both field beds and trial pots, and in many individuals flowering buds started to initiate. The plant, in fact, has two underground tubers, one of the previous year and second of the present year; the former being exploited by the plant in the proceeding year to sprout and attain vegetative growth. With the progress in vegetative growth, a small new tuber gets initiated in the close proximity of the previous year tuber, and this new tuber grows with the growing season. Ultimately, this newly produced tuber attains its critical size subjected to the favourable growth conditions in order to support the plant in the next sprouting season. In many trial pots at the KUBG, it was found that some of the vigorously growing individuals developed 3-4 underground tubers, instead of the usual 2, from the same plant during the single growing season. A few trials were conducted for the multiplication of this species by vegetative means from its underground tubers. In the subsequent experiments these tubers were cut vertically into 2, 3, 4, or even 5 sections depending upon the vigourness of the tuber and also each section gets a sizeable part of the apical bud. These sections were then planted in the trial pots at a depth of 2-3 cms buried under the soil. It was observed that in a few trial pots these vegetative sections developed into new plants. After repeated trials it was standardized that 2 sections in normal sized tubers and 3 to 4 sections in highly vigorous tubers are optimum for the establishment of such sections, and to successfully produce new plant. This observation got confirmed from the above trials; because once too many sections were made from the underground tuber, most of these sections did not get the critical portion of the apical bud essential for producing a new plant. Such a propagation trait of this species could be highly potentious for its rapid multiplication vegetatively.

Next in, the plant accessions of *Lagotis cashmeriana,* another high alpine target project species which were planted in the KUBG during the first year of the project, although sprouted quite satisfactorily, but showed stunted growth and ultimately dried up. The freshly collected accessions of this species during the second year supplemented with survived last year rhizomes when shifted to different experimental setups showed good results, although none entered into reproductive phase. This plant species has good prospects to propagate through vegetative means as in the natural habitats it does so through the same method.

Some of the individuals of another target project species, the *Aquilegia nivalis*, collected during the first year survived and produced flowers and seeds, although few in number; and the newly added plant accessions during the second year showed mixed results, a few of them withered while others sprouted afresh.

The field survey team was unable to collect *Hedysarum cachemirianum* from the collection sites mentioned by the previous workers, even after extensively surveying the mentioned sites during the first year. Fortunately, the species was collected from a relatively remote and floristically less-surveyed area of Kashmir Himalaya – Naranag. Only 14 accessions of the plant were collected during the present study. As a preplanned strategy practiced during the field surveys, the information regarding natural habitat conditions, such as soil, moisture level, slope, openness, shadiness, etc. for this species were noted down, so as to accordingly manipulate the conditions at the KUBG, to suit the transplanted plants of this species at least partially. The propagules of *Hedysarum cachemirianum* were planted in beds and watered occasionally, as the species needs moderate moisture conditions. Ten individuals survived and produced new twigs afresh in the same growing season. These individuals showed good response as most of them sprouted. In its natural habitat, the plant species possessed medium height associated with a robust, relatively long, thick and much-branched rhizome. These rhizomes possess many 1 cm long white-coloured vegetative buds. The latter produce new off-shoots in the natural habitats, which can be used for its macro-propagation in the *ex-situ* conditions.

Gentiana cachemirica, another interestingly bizarre plant species, inhabits typically gigantic rock outcrops in the alpines of the Kashmir Himalaya. The plant accessions of this species actually collected in the first year of the project had altogether failed, probably due to inadequate root-stock. After immense hard-work and exhaustive and highly laborious efforts in the successive collection trips during the second year, we succeeded in collecting a good number of fully rooted individuals or tufts of this species from its natural habitat. Typically, the plant clangs in the form of tufts within the narrow cracks on the surface of such rocks. On closer examining of such tufts, many individuals of the same plant can be found associated by their rhizomes. In such peculiar rhizomes, many individual plants are tightly intermingled with each other and it is very difficult, rather impossible, to separate them intact. This intertwined rhizomatous structure can be found in a thin layer of superficial soil inside the rock cracks. This rhizomatous structure then forces its entry by piercing deep into the rock crevices, creating space for itself. As a result of this situation, the survey team had to arrange hammers, pegs and other heavy tools, besides manpower to extract out the individuals of this species fully intact. In few cases, the research team succeeded in cracking the rock segments along the insertion area of the rhizome. It was found that the pressure of the rock had turned rhizome into a spreaded and flat (still intact) structure, with individual rhizomes squeezed sideways, appearing as if they have been hammered. Most portion of the rhizome was exclusively inside the rocks, with either no soil or a very thin layer of soil adhered to the main rhizome and its freshly formed roots/root hairs. It is because of this difficulty, only few entire rhizomatous structures (or intact plant tufts) could be collected while the rest were collected in the form of broken/fragmented rhizomes.

These propagules after planting in beds at the KUBG, survived for a short period of time in different experimental setups, except a few which were mostly in the form of intact ones. A few individuals in the latter case even produced flowers (1-2 flowers each) but failed to set seeds. During the spring season of the 2006, the previous year survived tufts started sprouting afresh in the form of very small, hard and stout, light-green rosettes of leaves.

Plant accessions of a highly restricted plant species of *Megacarpaea polyandra* showed relatively good vegetative growth and one individual even showed abnormal flowering (bolting), but the plant accessions added in the year 2005 showed poor results and only a few sprouted with no individual producing seeds. Majority of the transplanted individuals collected failed to sprout during 2006. It seems the root-stock appears to avoid too much of water as it gets rotted due to its softness. We apprehended that this transplanted material may have been rotted as a result of excessive water retention for quite a long period due to melting of overlying snow-cover. To verify this, we dug out some of the rhizomes and found them completely rotted. In fact, we were left with a few individuals of this species in trial pots at the KUBG. These survived individuals were obtained through the seed germination trials conducted in the previous year, and they attained appreciable vegetative growth.

Of all the target project species, the most narrow local endemic- *Meconopsis latifolia* – posed a serious problem vis-à-vis its establishment at the KUBG. The seedlings and adult individuals of this species, along with its rhizospheric soil, were brought from their natural habitat and planted in different experimental setups in the experimental plot. Initially, all the plant accessions survived but with the passage of time many of them withered away and just a single individual survived, and that too, showed very slow and stunted vegetative growth with no signs of reproduction. During the first year, we had obtained few seedlings of this species from the seed germination trials under the laboratory conditions, but when these seedlings were transplanted to soil mixed with sand in the trial pots, no one survived. We brought some amount of soil from its natural habitats and sowed some of the available seeds in it, so as to obviate the need for the transplantation. These seeds germinated and we were able to obtain few seedlings. As these seedlings started to gain height and produced the photosynthesizing leaves these seedlings along with petri-plates were directly transferred to the field, and put in complete shady conditions. To our bad luck, all these seedlings dried up just after one day.

During the month of May 2006, further 11 plant accessions of this species were added. A special bed established with small boulders crushed stones, sand and soil was prepared to mimic its natural habitat conditions for optimal survival and growth. In spite of this, only 6 to 7 individuals produced fresh leaves and appeared to have survived, while rest have withered away during the initial days.

In the beginning of June 2006, the planted propagules of all the 9 species sprouted in the field beds in varied numbers. These are *Gentiana cachemirica, Aquilegia nivalis, Hedysarum cachemirianum, Megacarpaea polyandra, Lagotis cashmeriana, Artemisia amygdalina, Saussurea costus, Aconitum kashmiricum,* and *Meconopsis latifolia* .

The propagation and maintenance of populations of these species under *ex situ* conditions seems to be better effected through their underground perrenating parts - roots/rhizomes- except in the *Meconopsis latifolia* in which the root/rhizome is too fragile to get established.

Practical relevance

The results achieved are of direct practical relevance for the conservation and sustainable use of plant biodiversity in this region. This will pave the way for the recovery, conservation and restoration of these economically important and critically endangered narrow endemic angiosperms of the Kashmir Himalaya through the provision of long-term backup collections for the sustained use by the local populace.

Constraints faced

★ The target species are local endemics, inhabiting very specialized microhabitats in sub alpine and alpine zones, often in very difficult and far off places. This, together with innate extreme rareness of these endemics, makes their location and collection very difficult; at times the survey team had to return empty handed even after exploring vast areas for days together.

★ The terrain inhabiting the target species is wholly mountainous and remains covered with snow usually from October to March, rendering it inaccessible for plant collection for about six months.

★ Due to their fragile nature and specialized microhabitat requirements, some of the target species, e.g. *Meconopsis latifolia* resist establishment in the KUBG.

★ Detailed tissue culture studies on the target species are time consuming and demand sufficient funding to be carried out to serve a useful purpose in *ex situ* conservation of these species.

Education and other promotional programmes

A useful programme was carried on at the High-altitude Botanical Garden, Gulmarg, to apprise a group of Kashmir University students about the role this

Botanical Garden can play in educating local people about importance of endemic plants and in promoting tourism of this State.

Conclusion

The present studies allow us to conclude that extreme habitat specialization in these nine species together with herbivory, human exploitation and inhospitable climatic conditions are the key factors restricting spatial spread and local abundance of the taxa. However, most the species are amenable to *ex situ* multiplication measures which could be exploited for restoration, maintenance and conservation of these species in the Kashmir Himalaya.

8. LITERATURE CITED

Balmford, A. and A. Long. 1994. Avian endemism and forest loss. *Nature* **372**: 623-624.

Crisp, M. D., S. Laffan, H. P. Linder, and A. Monro. 2001. Endemism in the Australian flora. *Journal of Biogeography* **28**: 183-198.

Dar, A. R., G. H. Dar and Zafer Reshi. 2006a. Conservation of *Artemisia amygdalina*-A Critically Endangered Endemic Plant Species of Kashmir Himalaya. *Endangered Species UPDATE* **23**: 34-39.

Dar, A. R., G. H. Dar and Zafer Reshi. 2006b. Recovery and restoration of some critically endangered endemic angiosperms of the Kashmir Himalaya. *Journal of Biological Sciences* **6**: 985-991.

Dar, G. H. and A. R. Naqshi. 1984. Some rare and little known Plants from Kashmir Himalaya. *Biol. Bull. India* **6:** 171 – 173.

Dar, G. H. and A. R. Naqshi. 2001. Threatened Flowering Plants of the Kashmir Himalaya – a checklist. *Oriental Science* **2**: 23 – 53.

Dar, G. H. and N. Aman. 2003. Endemic angiosperms of Kashmir: assessment and conservation, pp. 63 (abstract) *in*: National seminar on Recent Advances in Plant Science Research, October 12-14, 2003, Department of Botany, University of Kashmir, Srinagar, India.

Davis, S. D., V. H. Heywood, O. Herrera-MacBryde, J. Villa-Lobos, and A. C. Hamilton, editors. 1997. Centres of plant diversity. A guide and strategy for their conservation. Volume 3. The Americas. World Conservation Union Publication Unit, Cambridge, United Kingdom.

Dhar, U. 2002. Conservation implications of plant endemism in high-altitude Himalaya. *Current Science* **82**: 141-145.

Dhar, U. and P. Kachroo. 1983a. Alpine Flora of Kashmir Himalaya, Scientific Publishers, Jodhpur, India.

Dhar, U. and P. Kachroo. 1983b. Some remarkable features of endemism in Kashmir Himalaya. *In*: An Assessment of Threatened Plants of India (Eds. S. K. Jain and R. R. Rao) pp. 18-22. Proceedings of the Seminar held at Dehra Dun, 14-17 September, 1981. BSI, Howrah.

Hajra, P. K. 1983. Plants of Northwestern Himalayas with restricted distribution – a census. *In*: An Assessment of Threatened Plants of India (Eds. S. K. Jain and R. R. Rao) pp. 1 – 12. Proceedings of the Seminar held at Dehra Dun, 14 – 17 September, 1981. BSI, Howrah.

Kala, C. P. 1998. Ecology and Conservation of alpine meadows in the Valley of Flowers National Park, Garhwal Himalaya. Ph. D. thesis. Forest Research Institute, Dehra-dun, India.

Kala, C. P. 2000. Status and Conservation of Rare and Endangered Medicinal Plants of Indian Trans-Himalaya. Biological Conservation 93: 371-379.

Kapur, S. K. 1983. Threatened Medicinal Plants of Jammu and Kashmir. *J. Sci. Res. Plants. Med.* **4:** 40 – 46.

Kaul. M. K. 1997. Medicinal Plants of Kashmir and Ladakh, Temperate and Cold Arid Himalaya. Indus Publishing Co., New Delhi.

Kessler, M. 2000. Elevational gradients in species richness and endemism of selected plant groups in the central Bolivian Andes. *Plant Ecol.* **149**: 181-193

Kessler, M. 2001. Maximum plant-community endemism at intermediate intensities of anthropogenic disturbance in Bolivian montane forests. *Conservation Biology* **15**: 634-641.

Lamoreux, J. F., J. C. Morrison, T. H. Ricketts, D. M. Olson, E. Dinerstein, M. W. McKnight, and H. H. Shugart. 2006. Global tests of biodiversity concordance and the importance of endemism. *Nature* **440**: 212-214.

Leroux, S. J. and F. K. A. Schmiegelow. 2007. Biodiversity concordance and the importance of endemism. *Conservation Biology* **21**: 266-268.

Major, J. 1988. Endemism: a botanical perspective. Analytical biogeography (eds A. A. Myers and P. S. Giller), pp. 117-146. Chapman & Hall, London.

Maunder, M. 1988. Plants in Peril, 3. *Ulmus wallichiana* Planchon (Ulmaceae). *The Kew Magazine* **5**: 137 – 140.

Murashige, T; Skoog, F (1962). "A Revised Medium for Rapid Growth and Bio Assays with Tobacco Tissue Cultures". *Physiologia Plantarum*. **15** (3): 473–497.

Myers, N., R. A. Mittermeier, C. G. Mittermeier, G. A. B. Fonseca, and J. Kent. 2000. Biodiversity hotspots for conservation priorities. *Nature* **40**: 853-858.

Oostermeijer, J. G. B., S. H. Lujiten, and J. C. M. den Nijs. 2003. Integrating demographic and genetic approaches in Plant Conservation. Biological Conservation 113: 389-398.

Pitman, N. C. A. and P. M. Jorgensen. 2002. Estimating the size of the World's threatened flora. *Science* **298**: 989.

Saharia, V. B. 1982. Wildlife in India. Natraj Publishers, Dehra Dun. India.

Vergeer, P. R. Rangelink, A. Copal, and N. J. Ouborg. 2003. The interacting effects of genetic variation, habitat quality and population size on performance of *Succisa pratensis*. Journal of Ecology 91: 18-26.

Vir, Jee, P. Kachroo and U. Dhar. 1990. Endemism – a critical assessment. *In*: Advances in Frontier Areas of Plant Sciences (Eds. C. P. Malila, D. S. Bhatia).

Weekly, C. W., and T. Race. 2001. The breeding system of *Ziziphus celata* Judd and D. W. Hall (Rhamnaceae), a rare endemic plant of the Lake Wales Ridge, Florida, U. S. A: implications for recovery. Biological Conservation 100: 207-213.

Plate 1: Preparation of field and fixing of labels/sign boards

1.1 Beds prepared for planting the collected project plants

1.2 Project plot beds with fixed labels

1.3 Labeled project plot beds containing sprouted material of *Saussurea costus*

1.4 A view of project plot with sign board in front

Plate 2: Collected materials of some of the project species

2.1 *Saussurea costus* – Rhizomes

2.2 *Meconopsis latifolia* - Whole plant

2.3 *Megacarpaea polyandra* – whole plant with fruits

2.4 Prof. G. H. Dar (PI) examining *Megacarpaea polyandra* (fruits) and *Saussurea costus* (rhizomes)

Plate 3: Fruits / seeds of some of the project species

3.1 *Saussurea costus* - Seed heads

3.2 *Megacarpaea polyandra* – Seeded fruits

3.3 *Meconopsis latifolia*- Capsules with seeds

3.4 *Lagotis cashmeriana*- Seeds

Plate 4: Plants of the target species growing in KUBG

4.1 *Artemisia amygdalina*-veg. plant

4.2 *Megacarpaea polyandra* – veg. plant

4.3 *Saussurea costus* – vegetative plant

4.4 *Megacarpaea polyandra*- in flower

4.5 *Saussurea costus*- in flower

Plate 4: Continued

4.6 Sprouted specimens of *Aquilegia nivalis*

4.7 *Aquilegia nivalis* – vegetative plant

4.8 *Aquilegia nivalis* – in flower

4.9 *Aquilegia nivalis* – flower in closeup

4.10 *Aquilegia nivalis* - capsule

4.11 *Lagotis cashmeriana* – veg. plant

Plate 5: Seed biology experiments

5.1 Setting of seed biology experiments

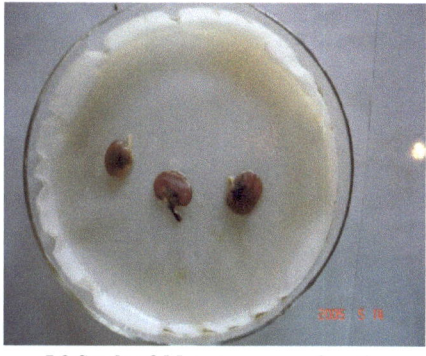

5.2 Seeds of *Megacarpaea polyandra* on germination trial

5.3 Geminating seeds of *Lagotis cashmeriana*

5.4 Germinating seeds of *Meconopsis latifolia*

5.5 Seedling of *Meconopsis latifolia*

Plate 6: Tissue culture experiments

6.1 Setting of tissue culture experiments

6.2 Seedling of *Megacarpaea polyandra* on culture medium

6.3 Seeds of *Lagotis cashmeriana* on culture medium

6.4 Seeds of *Meconopsis latifolia* on culture medium

6.5 Sprouted material (petioles) of *Aquilegia nivalis* on culture medium

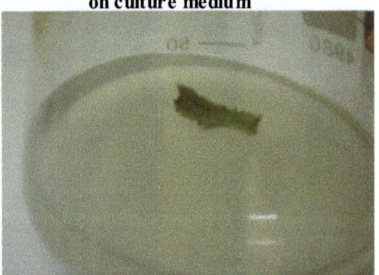

6.6 Sprouted material (leaf) of *Megacarpaea polyandra* on culture medium

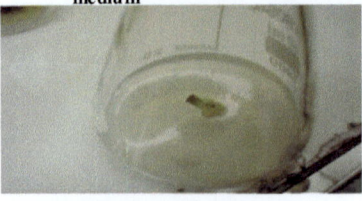

6.7 Callus formation (from leaf) in *Lagotis cahmeriana*